BUZZ

The Insect Files

Discovery Channel School
Science Collections

DISCOVERY CHANNEL SCHOOL™

1 2 3 4 5 6 7 8 9 10 PO 06 05 04 03 02 00

Discovery Communications, Inc., produces high-quality television programming,
interactive media, books, films, and consumer products. Discovery Networks, a division of Discovery
Communications Inc., operates and manages Discovery Channel, TLC, Animal Planet, Discovery Health Channel, and Travel Channel.

Writers: Jackie Ball, Vanessa Elder, Carolyn Jackson, Monique Peterson, Gary Raham, Rachel Roswal, Buffy Silverman, Alicia Slimmer, Anne Wright.
Editor: Anne Wright. **Photographs**: Cover, grasshopper, ©James Carmichael/Image Bank; pp. 2, 4–5, wasp, ©Dr. A. C. Twomey/Photo Researchers, Inc.; p. 8, moth eggs,
©ER Degginger/Photo Researchers, Inc.; grasshopper eggs, ©Dwight Kuhn; moth larvae, ©Kenneth M. Highfill/Photo Researchers, Inc.; pp. 8–9, grasshopper nymph and
adult, ©R. Simpson/Visuals Unlimited; p. 9, moth pupa, ©Gilbert Grant; p.9, adult moth, © M. H. Sharp/Photo Researchers, Inc.; pp. 10–11, cicada holes, ©Art Today; p. 12,
fly, ©Nuridsany et Perennou; cockraoch, ©Dee Breger; bee, ©Dr. Jerry Burgess, all Photo Researchers, Inc.; p. 13, silverfish, ©Oliver Meckes/Photo Researchers, Inc.;
diving beetle, ©Gary Meszaros/Visuals Unlimited; pp. 14 and 31, thorn bug, ©Valorie Hodgson/Visuals Unlimited; p. 15, wasp, ©Michael P. Gaoomski; hover fly, ©K. G.
Vock/Okapia/Photo Researchers, Inc.; caterpillar, ©Chuck Vaughn/Fremont, CA; beetle larvae, ©P. M. Choate, Entomology and Nematology Dept., University of Florida;
p. 16, fruit fly, ©Dwight Kuhn; "Germander Speedwell and Ladybug," ©The Pierpont Morgan Library/Art Resource; p. 17, Egyptian scarab, ©The Granger Collection; dung
beetle, ©Robert J. Ellison/Photo Researchers, Inc.; hummingbird moth, ©Joe McDonald/Visuals Unlimited; pp. 18–19, soldier termite, ©John D Cunningham; worker termite; ©George Loun; fungus garden, ©Don W. Fawcett; termite queen ©Kjell B. Sandved, all Visuals Unlimited; termite mound, Kenya, ©J. Ferner/Visuals Unlimited; p. 20
Berta Vogel Scharrer, ©Ted Burrows; p. 21, woodroach, ©G & C Merker/Visuals Unlimited; p. 23 locust, ©Don W. Fawcett, mosquito, ©Richard Walters/Visuals Unlimited;
p. 24, army ants ©Leonide Principe/Photo Researchers, Inc.; p. 25, army ants, ©Rudolph Freund/Photo Researchers, Inc.; p. 27, flea shrimp fossil, ©Louise K Broman;
prawn fossil, ©E. R. Degginger; crustacean fossils, ©Joyce Photographics; sea scorpion fossil, ©Kaj R. Svensson, all Photo Researchers, Inc.; pp. 28–29 Palos Verdes blue
butterfly, Courtesy of Arthur Bonner & Palos Verdes Blue Butterfly Reserve; photo of Arthur Bonner, by Axel Koester/Corbis Sygma; p. 31 giant male water bug, ©Ken
Lucas, Ecuadorian caterpillar, ©Davis Pearson, from Visuals Unlimited; stink bug, ©1997 Alan and Linda Detrick/Photo Researchers, Inc.; all other photographs are
©COREL. "Leiningen and the Ants," by Carl Stephenson in GHOSTLY, GHOULISH, GRIPPING TALES, Franklin Watts, New York, ©1983. Reprinted with permission.

CONTENTS

BUZZ

Swarming Multitudes

Buzz ... hiss ... chirp ... A fly, a snake, and a bird? Guess again. They're *all* insects. In terms of the number of species, insects are tops—hands down. There are more known species of insects than all other animals put together. What's more, scientists estimate that there are millions more species they haven't even identified.

What are the secrets of insects' success? One is the ability to fly, which helps many species of insects evade their natural enemies and travel long distances in search of food or a mate. Another is their body structure, something unique to insects.

Where do they live? Other than the middle of the ocean, everywhere— cold climates and hot ones, on mountaintops, and in lakes and rivers. Over the 350 million years they've been around, insects have adapted to a great variety of environments, developing defenses to evade predators, as well as feeding mechanisms that enable them to obtain food almost anywhere.

Want the details? In BUZZ, Discovery Channel looks at this amazing world of insects. There is a lot of great stuff to learn about all those little creatures.

The Insect Files

Insects ... 4

At-A-Glance
What do 900,000 different kinds of insects have in common?

When does this cicada come out to play? See page 11.

This butterfly is an inspiration to the auto industry. See page 30.

Final Project

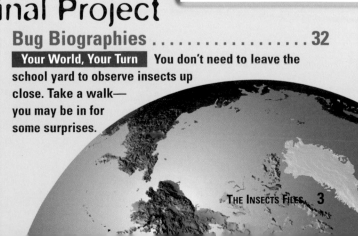

Insects

Ants at a picnic, a fly on the wall . . . Just look out the window or go out in the yard—you're bound to see insects. They're around us all the time. Sometimes we're aware of them, other times we're not. But they are there—in the house, flying through the air, in the puddle at the end of the driveway, underground, up a tree.

And it's a good thing. Sure, some insects are pests—they can destroy crops and spread disease, but we also rely on them to pollinate fruit trees, control insect pests, and dispose of waste. Many birds and other animals depend on insects for food.

To be an insect, there are a few requirements: three body segments and six legs. The body parts are the head, the thorax, and the abdomen. Many insects also share features, such as wings and antennae. Like spiders, shellfish, ticks, and certain other animals, adult insects are covered by a protective external skeleton. With a wasp as your guide, take a closer look at just what it takes to be an insect.

FEATURE: ANTENNAE

If you know antennae as "feelers," you can guess one of their functions. Most insects use them for touch, but they can also use antennae for smell and taste and sometimes to "hear" and to grab prey. (Wasps use them mainly for smell and touch.)

FEATURE: EYES

Most insects have two kinds of eyes. There are small, "simple" eyes, which are only sensitive to light, and "compound" eyes made up of many different single eyes. Scientists think compound eyes see light and detect movement in all directions but cannot make the fine visual distinctions humans can.

REQUIREMENT: EXOSKELETON

The exoskeleton is composed mainly of a tough, flexible substance called chitin. Unlike our skeletons, exoskeletons don't grow, so as insects get larger, they shed their small exoskeletons and grow new, larger ones.

REQUIREMENT: HEAD
Body parts necessary for feeding, sight, smell, and touch—the mouth, antennae, and eyes—are located on the head.

FEATURE: WINGS
Wasps, like many insects, have wings. Wasps have two pairs. They are hooked together. Insects' wings are flexible, which makes flying easy. Wings may serve other functions, too: aiding in camouflage, helping members of the same species identify each other, or absorbing light, which helps an insect regulate body temperature.

REQUIREMENT: ABDOMEN
Within the abdomen are the digestive tract and the reproductive system. Like the thorax, the abdomen also contains spiracles.

REQUIREMENT: THORAX
Wings and legs are attached to this part of the body. It's also one of the places where breathing occurs. Air comes in through small holes along the thorax called spiracles.

REQUIREMENT: LEGS
All insects have six legs. At the end of the legs are claws, designed for holding onto surfaces and prey and for digging.

Swoop and Scoop

Talking with a dragonfly

Q: You're a dragonfly—colorful, graceful, one of the most beautiful types of flies on Earth.
A: Why, thank you. But I'm really not.

Q: Not colorful, graceful, and beautiful?
A: Oh, I am all those things—and so much more. I am a descendant of the world's oldest and largest insects. Dragonflies as big as seagulls were gliding above Earth 300 million years ago—before the dinosaurs. And even today we're among the world's largest insects. Some kinds of dragonflies have wingspans twice as wide as the palm of your hand.

Q: Very impressive. But what are you not?
A: I am not a type of fly. Don't get me wrong—I have nothing against houseflies and mosquitoes and such. In fact, I can eat them by the dozen.

Q: What are some ways you're different?
A: Let's start with the basics: wings. My four are, like the rest of me, very attractive, with delicate veins forming lacy patterns. They're arranged in two neat sets, one behind the other. I alternate the sets when I fly, first front, then back, nice and smooth, a noiseless 30 beats a second.

Q: And flies?
A: They have only two wings, with a set of stubby things behind them to help them balance. No wonder they're always frantically flapping around and making that annoying buzzing sound.

Q: But at least flies can fold their wings back. You must get tired of holding yours out horizontally all the time.
A: Big deal. Who needs folding wings? My fixed wings work phenomenally well—so well that they haven't changed for millions of years. My wing muscles are stronger than any other insect's, so I am an absolute wizard in the air. I can hover, fly backward and upside down. I can fly far, and I can fly as fast as a deer can run. Besides, having my wings stretched out at all times means I'm always ready to soar, sail, and do my favorite thing—swoop.

Q: Why is swooping your favorite thing?
A: Because that's how I catch my food. Of course, before I catch it, I have to spot it. But because of my amazing eyes, that's pretty easy. Which is another way I'm different from your ordinary housefly.

Q: Wait a second. Flies have large eyes made up of lots of little surfaces that fit together, just like yours.
A: What you're describing are compound eyes, which most insects have—including dragonflies, and yes, true flies. But a fly's eyes are NOT just like mine. Mine are much bigger relative to my head than a fly's. And they're made up of as many as 30,000 facets or surfaces. Houseflies have only 4,000 or so. No contest.

Q: OK. But what's the advantage of having more facets?
A: The more the merrier! The more facets, the better I see. Some facets point up, some down, some forward, some backward. I can see in all directions—a perfect circle of sight! I can also see up to 40 feet (12 m) away. If there's anything to eat, I'll spot it. If there's anything that wants to eat me, I'll escape.

Q: After you spot supper, what do you do?
A: I lock my strong, bristly legs together so they form a kind of basket. Then I swoop down, scoop up the gnat or mosquito or whatever, and presto! The picnic's in the basket. I munch while I fly.

Inflight dining at its best.

Q: Sounds like fun.
A: It is, it is. Too bad life's so short—that is, as an adult. Grown-up dragonflies live only a few weeks. But in our early stage, as nymphs, we live a year or two. And that stage is fun, too. That's when I really developed my hunting skills, even though it was in a whole different environment.

Q: What do you mean?
A: As a nymph I was aquatic, not aerial. I lived in pond water and ate small aquatic creatures—on a good day, even small fish. Nymphs have a special weapon: a kind of folding mask half as long as their whole body, which shoots out to capture prey. It has hooks and spikes on the end to grasp and pull the prey back into the nymph's mouth. See, even in those days I was scooping up my supper. And even then I was fast! I could shoot out that mask and reel in my prize about ten times a second.

Q: So could you sum up the secrets of your species' long, successful life?
A: Sure. Size, sight, speed, and unsurpassed skills in the sky. And, of course, modesty.

Activity

FLIGHT DATA Lots of insects fly, but wing size, shape, and number may vary from species to species. Flight speeds and wing beats per second also vary by species. A horsefly was clocked at about 90 miles an hour. A tiny midge flaps its wings more than 1,000 times a second while honeybees beat their wings 190 times a second. Find out the flight speed and wing beats per second for 10 different kinds of insects. Plot the information on a graph. What can you conclude about the relation of wing beats per second and flight speeds?

MY, HOW YOU'VE

Every insect starts life as an egg. The change from egg to mature adult is called metamorphosis, which means "shape change." Some types of insects undergo complete metamorphosis—the insect is different at each stage of development. In other types of insects, metamorphosis is incomplete, meaning that the insect starts out resembling a small adult, and, as it ages, gets bigger.

Complete Metamorphosis: Female Moth

Day 1

EGG

A female cecropia moth can lay more than 100 eggs. With a glue that its body produces, the moth attaches the eggs in small groups to the undersides of leaves.

Days 7–14

LARVA

The egg hatches and a larva, or caterpillar, emerges. It is an insect, even with a hairy body and more than six legs. Its main function is eating. As it grows, it will shed several exoskeletons and its color will change from black to green with orange knobs.

Incomplete Metamorphosis: Female Grasshopper

Day 1

EGG

In late summer or early fall, a female *Romalea microptera* (grasshopper) deposits about 20 to 100 fertilized eggs in the soil. Within each egg, an embryo starts to develop, then stops and becomes dormant over the cold winter months. Once the temperature rises in the spring, development begins again.

7–9 Months

NYMPHS

Each embryo pushes its way to the surface. The nymph is pale, but within hours its color will be the same as an adult's. The young nymph has small pads where wings will grow. It eats grass, leaves, and other vegetation. As it grows, it will shed its exoskeleton several times, each time looking more and more like an adult.

CHANGED

Moth
Hyulophora cecropia

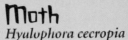

2-3 Months

PUPA

The full-grown cater-pillar moves close to the ground and starts spinning its silk cocoon. It takes several days to complete. The finished cocoon will be tough and weather resist-ant. Over the winter, the pupa rests. As spring approaches, the change begins.

9 Months

ADULT

In late spring, the full-grown moth emerges from its cocoon. It will make pumping motions to force blood into its limp wings and straighten them out. After a day, the moth will begin to fly. During its two-week life, the moth mates and the cycle starts all over again.

30–40 days later

ADULT

After the nymph has shed and grown about five exoskeletons, it is a full-grown adult. It has two pairs of fully developed wings and is capable of reproducing. It will live a few more months.

Grasshopper
Romalea microptera

Activity

LIFESTYLES Investigate the life cycles of these types of insects:
Fireflies
Click beetles
Katydids
Walkingsticks
Waterbugs
Carpenter ants
Hornets
Damselflies

Answer the following questions about each: What type of metamorphosis does each undergo? Where are the eggs laid? What are the young insects called? Where do the young insects live? What do the young insects eat? Use the answers to make a data base. Sort the data to look for patterns.

WE ARE THE WORLD

Except for the middle of the ocean, insects live everywhere. No matter where they are, all insects share an ancestry and physical characteristics such as three body parts, six legs, and breathing mechanisms. Yet there are more different species of insects than any other animals on Earth. Why? Because insects have developed many different adaptations that enable various species to survive in almost any type of environment. Some species are adapted to live in the hot, arid climates, while others require cold temperatures. In between are hundreds of thousands of types of insects in a wide range of habitats.

Insects Rule

There are over 1.4 million species of animals known to scientists. Of these, about 900,000 species are insects. Percentage-wise, that means that insects comprise about 64 percent of all animal species put together.

All other animal species 36%

Insects 64%

Ahead of the Pack

How do insects stack up against their own kind? Compare them to their fellow invertebrates—spiders, worms, clams, starfish, and crabs—and you'll find insects on top. There are almost four times as many insect species as all other invertebrates combined.

Other invertebrates 20%

Insects 80%

Tropical rose chafer

Beetle Mania

There are more than 250,000 species of beetles. That's more than any other group of insect! While beetles dominate (in numbers), there are some other big groups as well that, with beetles, account for about two-thirds of the known insect species.

Beetles: more than 250,000 species (2.5×10^5)

Butterflies and Moths: more than 150,000 species (1.5×10^5)

Bees, Ants, and Wasps: 108,000 species (1.08×10^5)

True Bugs: more than 82,000 species (8.2×10^4)

Grasshoppers and Crickets: 23,000 species (2.3×10^4)

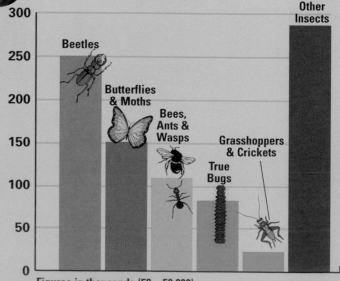

Beetles

Butterflies & Moths

Bees, Ants & Wasps

True Bugs

Grasshoppers & Crickets

Other Insects

Figures in thousands (50 = 50,000)

INSECT CENSUS

We know there are many different species of insects, but what about individual insects? Well, no one could actually count every insect in the world. Canadian biologist Brian Hocking estimated that there are 1 quintillion insects on Earth. That's **1,000,000,000,000,000,000** (1×10^{18})!

Together, those insects would weigh about 2.7 billion tons. The 6 billion people now living on Earth weigh about 4.5 hundred million tons. The total insect weight is about 6 times that of the total human weight. To put it in perspective, a 150-pound human weighs about 600 times more than the Goliath beetle, the heaviest known insect, which weighs less than 4 ounces!

Safety in Numbers

Cicadas are often called locusts, but they are not. Unlike locusts, cicadas cause relatively little damage to plant life. The nymphs develop underground over a period of 13 or 17 years, depending on the type of cicada. Then they emerge all at once, and when they do, cicadas can number about one million per acre. Scientists speculate that their large numbers help ensure their survival. They have many natural enemies—birds, dogs, raccoons, and other animals—and few defenses, but because they are so numerous, there are sure to be enough survivors to reproduce. Each female cicada will lay about 600 eggs during the month she lives above ground.

Cicadas emerging from their holes

Looking Ahead

Scientists don't know how many more species of animals are yet to be discovered. But biologists agree that most of the undiscovered animals are probably insects—perhaps as many as 30 million species. Huge numbers of undiscovered species live in tropical forests in Latin America. Many will never be discovered because people are destroying their rain forest environments.

Activity

MAKING A HEAD COUNT Conduct an insect census. Choose an insect that is normally found in your backyard or on the school grounds and try to estimate the population of the insect there. You'll never be able to count them all. You'll have to extrapolate, which means counting a small number and estimating what percentage of the total this represents. Describe the method you used to extrapolate and share your results with the class. Why would it be important to know the insect population of a particular region?

FOOD STUFF

The saying "you are what you eat" is especially true of insects. One way to distinguish among different types of insects is by their mouthparts. Mosquitoes have piercing and sucking mouthparts that enable them to bite and drink blood, adult butterflies have long tongues that they use to drink nectar from flowers, and grasshoppers have powerful jaws to bite and chew plant matter. Over time insects have adapted to many environments. With their specialized mouthparts, insects can eat food ranging from paste to clothing to hair to tobacco. But within each species, the diet is usually limited to a few foods. The limited diets mean that many types of insects can live together within one ecosystem because they do not compete for food.

In addition to specialized mouthparts, the digestive systems of many types of insects are adapted to particular foods. Termites have a microorganism in their intestines that helps to digest the wood they eat. Houseflies, which eat only liquids, can liquefy solid foods with saliva.

Check out the dietary data on some familiar insects.

Fly mouth magnified about 40 times

Waiter, There's a Fly in My Soup

Why would a fly feed on soup? The housefly doesn't have much choice. It has a special, sponge-like mouthpart (above, right) designed to "eat" only liquids. But another adaptation allows flies to eat a variety of foods, enabling them to survive in a range of environments. When a fly dips into solid foods, like sugar, it actually liquifies the food first by dissolving it with its saliva. Then it laps up the liquid.

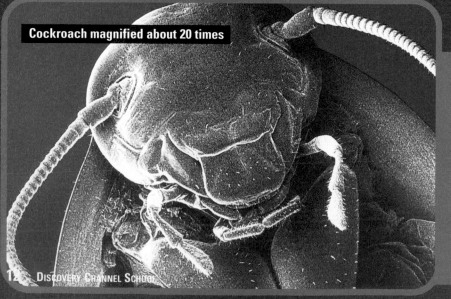

Cockroach magnified about 20 times

Varied Diet

After surviving over 300 million years on Earth, cockroaches are still going strong. Today there are more than 4,000 species of them. One reason for their success is their powerful jaws, which move side-to-side. Some species can use their jaws to eat almost anything—plants, animals, paper, and even clothing so they can always find something to eat in human households.

Honeybee magnified 3 times

Sweet Tooth

The adult bee uses its long, pointed tongue to suck up nectar from flowers. The nectar goes into a storage pouch in the bee's body. The bee also collects pollen from flowers to feed to larvae. As the bees feed, some of the pollen sticks to their legs and which gets transferred from one flower to another. That enables the flower to reproduce.

Honeybees have very specific feeding habits. They have a well-developed sense of taste and will fly several miles from their nest to find just the right kind of flower. And once they find a good feeding spot, they return to the hive to tell the other bees. A honeybee communicates the location of a patch of flowers with a dance. The dance is a kind of map to the spot.

The bees return to their hive laden with food. There, they regurgitate the nectar to other bees. Those bees, in turn, regurgitate it again and mix it with pollen to form "bee bread," to feed to the larvae, or deposit into cells. That substance in the cells is what we know as honey.

Book Lovers

The tiny silverfish's diet consists of sugars, starches, and proteins. In the modern world, it meets its dietary needs by eating paste, glue, and sizing, a chemical used on paper. Its flat body enables it to slide into bookbindings and behind peeling wallpaper in search of a meal.

Silverfish magnified about 30 times

Food Facts

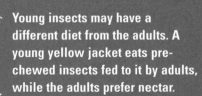

- Young insects may have a different diet from the adults. A young yellow jacket eats pre-chewed insects fed to it by adults, while the adults prefer nectar.

- Many insects don't drink water. They get the liquid they need from the food they eat.

- The diving beetle (right) eats tadpoles and small fish, as well as other insects.

- Many types of ants feed on "honeydew," not the fruit, but a sweet substance made by aphids, pest insects that feed on plants. Some ants even build shelters to protect their aphids.

Diving beetle magnified about 3 times

DRESSING FOR

An insect's life is brief and risky. Birds, mice, lizards, frogs, spiders, bears—hungry predators lurk everywhere. What's an insect to do? Fly away? Not all insects can fly, and even if they can, predators may fly faster. Fortunately, there are other lines of defense. Some insects hide out by blending in with their surroundings. Others fool predators into thinking they're a thorn or bird droppings. Many poisonous or stinging insects have brightly colored wings or bodies that warn enemies to stay away. And some insects can look more dangerous than they really are.

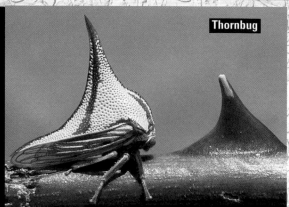
Thornbug

Katydid Leaf Me Alone

Members of the grasshopper family, katydids are known for their nighttime singing, but they also look interesting. They can evade predators because their wings resemble leaves on trees and shrubs. The wings feature a vein pattern that looks similar to leaves and small brown spots like the diseased areas and insect nibblings found on many leaves.

Praying Mantis
Best Costume, Best Actor

A green mantis rests on leaves or in tall grasses; a brown one hides on tree bark; and a pink mantis, a kind of Asian mantis, stays hidden among flowers that bloom where it lives. Mantises use their camouflage offensively and defensively. They prey on other insects. An insect will crawl close to a camouflaged, unmoving mantis without noticing it. Then, when the insect is within reach, the mantis snags it with its strong front legs and devours it.

Io Moth The Evil Eye

Like most moths, an io moth is active at night and rests during the day. Its front wings are hard to distinguish from the tree bark on which the moth rests. But occasionally a sharp-eyed bird spots it anyway. The moth, sensing a predator is near, lifts its front wings to show bright eyespots on its hind wings. The startled bird sees what looks like an owl's huge eyes. Before the bird realizes it has been tricked, the moth might have time to escape.

SUCCESS

Can you tell these two insects apart?
One is a stinging wasp. The other is a harmless hover fly.

Yellow Jacket Wasp The Real Thing

A yellow jacket wasp (upper right) raises its young in underground nests. If an animal approaches the nest, the adults swarm out and sting the intruder. Once stung, predators know that bright colors and loud buzzing mean "stay away!"

Hover Fly The Great Pretender

A hover fly (lower right) isn't poisonous and doesn't sting. But potential predators are often fooled into thinking it's a wasp. The fly's black-and-yellow stripes make it look like a yellow jacket. And a hover fly can mimic a wasp's long antennae by waving its front legs above its head. It even wags its wings to imitate the rocking motion wasps make when feeding. As a last-ditch effort, a hover fly caught in a bird's beak can make the same sound as a wasp.

Swallowtail Caterpillar and Tortoise Beetle Larvae Looks Aren't Everything

A swallowtail caterpillar (above, left) looks like bird droppings, and that's a plus. Not surprisingly, birds and other predators see a swallowtail caterpillar and fly on by in search of something that looks more appetizing.

Tortoise beetle larvae (above, right) go for the same look by picking up their own droppings, and other debris, and sticking them on the sharp spines along their backs. The effect on predators is the same.

Activity

CAMOUFLAGE Insects adapt their coloring to their environments. The adaptation is gradual and happens over many generations. Try this experiment to see how it works. You'll need two or three bags of M&Ms®, a bowl, and sheets of blue, green, or red paper.

1) Empty the M&Ms into the bowl. This is intended to model the "genetic pool," i. e., all the possible colors.

2) Grab a handful of M&Ms from the bowl. Place them on a sheet of paper. This represents a generation.

3) Remove about a third of the M&Ms, selecting those with the most contrasting colors. This simulates a predator-prey situation, where predators pick the most "obvious" prey in an environment.

4) Place the remaining M&Ms back in the bowl. Even after one generation, the gene pool will begin to show a bias toward the color that matches the paper.

5) Repeat steps 2–4 for several more generations.

After several generations, the M&Ms remaining on the paper will be the same color as the paper. With your help, the M&Ms have "adapted" to their environment.

DOING GOOD WORKS.

Real Ladies

Many insects behave in ways that benefit humans. Bees make honey and wax, and "silk worms," which are actually caterpillars, make silk, of course. Insects searching for nectar carry pollen from one flower to another. Some species prey on insects that harm crops or transmit disease. Others turn and fertilize soil as they dig to build nests or bury waste. And still others teach scientists a lot about human beings.

Monks in the Middle Ages noticed that ladybugs ate the aphids that were destroying grapevines. Ever since, these members of the beetle family have been famous for their enormous appetite. During its lifetime, one ladybug can consume more than 5,000 aphids, plant lice that attack fruit trees and vines. And even the young get into the act: A larva can eat 300 aphids in two weeks. Ladybugs also feed on scale insects, spider mites, mealybugs, and other pests. Many of today's farmers use ladybugs instead of chemical pesticides to grow bug-free crops.

A 16th-century manuscript shows a ladybug.

Getting Inside the Problem

When you hear the word "wasp," you probably think sting. And you're right, some wasps do indeed sting. But there are other wasps that not only don't sting, they help get rid of insect pests—flies, aphids, and caterpillars that feed on crops and trees. These parasitic wasps deposit eggs in pests. The wasp larvae feed on their victims, ultimately destroying them.

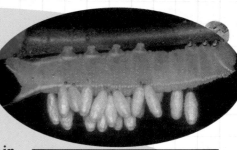

Wasp eggs attached to a caterpillar

SUPER FLIES

Fruit flies can destroy whole crops. But they have certain characteristics that scientists appreciate. For nearly a century scientists have been using the tiny fruit fly to study patterns of human heredity. Why the fruit fly? Well, it is not difficult to maintain in a lab and has a simple genetic structure, which makes it easy for scientists to interpret the results of their experiments. Fruit flies also reproduce quickly (it takes about two weeks to produce a new generation), so experiments can be conducted in a very short time.

IT'S A BIRD, IT'S A PLANE . . . IT'S A MOTH

The hummingbird moth (bottom left) may be confused with its namesake, the hummingbird (top). They look alike, sound alike, and both eat nectar with long tongues. Pollen from the flowers they feed on sticks to their wings or bodies. When they feed again, some of that pollen is transferred to other flowers, enabling the flowers to reproduce. Unlike the hummingbird, this moth does its work at night, pollinating flowers that are visible in the dark.

THE DIRTY WORK

"[T]hey are there in their hundreds, large and small, of every sort, shape and size, hastening to carve themselves a slice of the common cake."
J. Henri Fabré, 19th-century French entomologist and writer

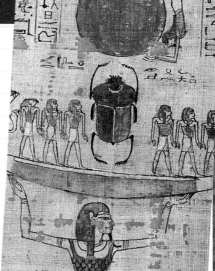

Cake? Not exactly. The writer was describing dung beetles. Found on every continent except Antarctica, these insects help the environment. As the name suggests, they deal in dung, moving it underground where it fertilizes the soil. They also eat dung and, by digesting it, speed up the process of converting the manure to fertilizer. What's more, flies are attracted to dung, so when dung beetles get rid of it, the flies disappear, too.

Within minutes after a pat of dung is dropped, the beetles go to work. Using their specially shaped heads as scoops and their antennae as shapers, they form the dung into large balls, which they roll to underground nests. In the nests, female beetles lay their eggs in the manure. The larvae then live off the fecal matter, devouring it and the disease-spreading parasitic worms and maggots that live in it. We're left with better soil for our plants and cleaner ground to walk on.

The dung beetle was sacred to the ancient Egyptians. It symbolized regeneration and renewal. The beetle, which the Egyptians called a scarab, was so important that when a pharaoh died, his heart was removed and replaced by a scarab carved of stone.

Activity

INSECT ATTRACTION Gardeners appreciate the work of pollinating insects. They often plant particular flowers to try to lure pollinators to their gardens. Find out what types of butterflies and moths are common where you live. Design a garden that will attract them.

HIGH SOCIETY: A TERMITE METROPOLIS

S UDDENLY YOU'VE SHRUNK! You find yourself standing before a termite tower in Kenya. It's huge—it stands over you like a skyscraper! Or maybe a city that's home to some 5 million termites. The termites range in age and each one performs a particular role that contributes to the survival of the whole community. Time to explore!

Termites are blind, but they have a great sense of smell, so you'd better smell like one of them to wander unnoticed through their tower. You have a vial of a pheromone, a chemical produced by termites. You spray yourself with some of it and you smell just like a termite. Time to head for the tower.

Fitting In

You pass some mushrooms. Why mushrooms? You spot some termites—workers reinforcing the outer walls with wads of chewed-up wood, dried plant matter, and feces. This creates a more secure shelter for the termites and speeds up the decomposition process. You move close to the wall and reach out and touch it. It's as hard as a rock. You spot an opening. The workers don't notice you—the pheromone must be working. But wait—a soldier termite is looking your way. You know she's a soldier because her head is much bigger than a worker's. It holds a huge pair of sawtoothed jaws. She swings her head toward you, massive jaws open . . . and walks on by. Phew! You hurry into the tower. Termites are everywhere—some workers are carrying scraps of food and others are offering regurgitated food to the soldiers, who can't feed themselves.

The Inside Story

You notice how warm it is when, all of a sudden, you feel a breeze. In termite towers cool air moves up from the lower part of the tower, out a central chimney, then back down again where it is cooled by air vents in the walls. You judge the temperature inside is around 85°F.

Soldier termites use their large jaws to attack their enemies.

Worker termites are the smallest adult termites in the mound.

Termites cultivate fungus gardens for food.

The queen termite's body is swollen with eggs.

Time to visit the rooms. You start to climb. First stop looks sort of like a garden. The crop seems to be fungus (something these termites like to eat). It's fairly humid, so the garden is thriving. But where did they get the "seeds" to plant the fungus? Now you remember those mushrooms. That's it! Workers must gather the spores, or reproductive bodies, shed by the mushrooms and "plant" them inside.

Where are those workers going? You follow them out of the garden and up inside the tower. They lead you to a brood chamber, a room filled with a squirming mass of pale nymphs fed by the workers.

Royal Visit

The next place you visit is the royal cell. You head up higher, but all you see are large openings. You figure you must have passed it so you head back down. Then you see soldiers guarding an entrance. They threaten and start to shove, but eventually let you pass. Your first glimpse of the queen is not pretty. She's huge, much bigger than the other termites, and shaped like a hot dog. Workers feed her constantly. Her abdomen bulges with eggs, which emerge steadily. She'll live for up to 50 years.

Quick Exit

Suddenly the chamber shakes violently. Dirt drops from the ceiling. You hurry through the tunnels, which are filling up with dirt and hordes of running workers. You spy an exit, but it's blocked by a giant paw. You slide out and start to run. Turning to see what's going on, you spot an aardvark pawing its way into the mound in search of food. You quickly grow to normal size and shoo the aardvark away. Take one last look at the tower: Some worker termites are already repairing the wall. No wonder they call them workers!

Activity

ANT-ICS Termites are social insects; they live in groups and perform specific duties within the group. Ants and some species of bees and wasps are social insects, too. Find an ant hill near your school or home. Observe it for several days and keep a record of what you see. Do you see ants working on the ant hill itself or bringing in food for the others? When they leave the ant hill, do they move in any particular patterns? Try leaving a few breadcrumbs nearby and see if they move toward them and pick them up.

Making
Cockroaches
a career

What can the cockroach brain teach us about the human brain? Dr. Berta Scharrer devoted much of her career to finding out. Her work with cockroaches led to some important discoveries.

From an early age, Berta Vogel wanted to become a scientist even though she knew that a scientific career for a woman would not be easy. She enrolled in the University of Munich, in her native Germany, and completed her doctorate in 1930. At the university, she met her future husband, Ernst Scharrer. His study of fish revealed that certain nerve cells in fish brains produced a substance that other scientists associated only with another part of animal bodies. Berta Scharrer found evidence of the same type of nerve cell activity in invertebrates. Since their findings were contrary to what most scientists thought, they decided to conduct comparative studies: Berta Scharrer would investigate the invertebrates and her husband the vertebrates.

The Resourceful Researcher

Within a few years, as the outbreak of World War II approached, the Scharrers left Germany because they objected to the policies of the Nazi government toward their Jewish colleagues. They came to the University of Chicago, but they had to leave their research in Germany. To make matters worse, Berta Scharrer had no money for lab materials and was limited to studying the only invertebrate specimens available—small flies.

One day, a university custodian offered her an alternative—a cockroach from the basement. Berta realized that the cockroach was much better suited to her work. She began setting traps and collecting the insects herself.

When her husband got an appointment at Rockefeller Institute in New York, Scharrer moved with him. She was eager to continue her research but was disappointed to find that the new lab was roach free.

Prospects looked grim until a crate full of monkeys arrived from South America. In the bottom were plenty of Madeira cockroaches, also known as South American woodroaches. Scharrer fed the slow-moving, 2-inch-long insects fruit and dog food. She nurtured a breeding colony that would thrive under her watchful eye for the next five decades.

Cockroach Behavior

Scharrer determined how to pair particular male and female cockroaches to promote reproduction. In later years, she discovered that introducing a roach that had just undergone a surgical experiment to another of the opposite sex actually improved the insect's recovery rate. "Just by trial and error, you hit on certain things," she noted.

A Closer Look

To study their brains and nervous systems directly, Scharrer performed microsurgery on the roaches. She was able to isolate and manipulate various brain parts of mature and immature cockroaches. She discovered that a substance produced by the nerve cells had a direct effect on the roaches'

South American woodroach

hormonal balance and metamorphosis. When Scharrer removed specific parts of the cockroach brain, roach nymphs went through premature growth cycles, and the eggs in adult females did not develop properly.

Later Work

The medical and scientific communities benefited from the work of Scharrer and her colleagues in a number of ways: Their findings contributed to the understanding of early development and growth in humans and other animals. Their research was a factor in the discovery of communication between the nervous system and the immune system. Her experiments with roaches led scientists to determine that the brain releases natural painkillers. Doctors have used this knowledge to help patients recover faster after surgery.

Cockroach Qualifications

What made the cockroaches such good research subjects? Certainly not the fact that they give off an offensive odor when disturbed. But the good qualities outweighed the bad. Their large size made it easy to observe what was happening inside them. They moved slowly and were not hard to handle. Because the roaches lived for about two and a half years, scientists could observe the results of experiments over a long time. The fact that they give birth to live young impressed Scharrer because it meant that she didn't have to wait for eggs to hatch and young roaches to develop. She wrote, "At first, they're little white nymphs, but they're ready to go about their affairs." All in all, the chance finding in a lab basement yielded a wealth of scientific knowledge.

Literary Life

Cockroaches are found in literature as well as labs. In the novel **Metamorphosis**, by Austrian-Czech writer **Franz Kafka,** the main character is transformed overnight from a human being to an insect that fits the description of a cockroach.

Activity

MAKING A HOME Think of a type of insect you would like to study. How would you go about setting up a colony in your lab? Where would you find the insects? Do some reading to figure out the best way to house the insects. Be sure to look at temperature, light, and humidity levels. What would you feed them and how often? Do they need water? Write up your findings and include a sketch of what a cage would look like.

Don't Bug Me, PLEASE!

Insects are everywhere. But different species develop regionally based on food availability, predators, and other conditions. They bite, sting, and destroy crops to get food and protect themselves. Sometimes humans get caught in the middle. Look at the map. You will see where insects and humans collide.

North America

South America

❶ **Colorado Potato Beetle, Western U.S.** Before the mid-19th century, the Colorado potato beetle ate native weeds. But when settlers in the West started cultivating potatoes, the beetles switched to this new food source. They ate their way east, even across the Atlantic to Europe. This insect quickly develops resistance to insecticides so other methods of control, such as handpicking and the use of natural predators, are the most effective ways to combat the pest.

❷ **Boll Weevil, Texas and Mexico** In the mid-19th century, the boll weevil virtually wiped out Mexico's cotton crop. Forty years later, the pest made its way north to Texas, where it became the most serious threat ever to U.S. cotton crops. Boll weevils are effective destroyers partly because they reproduce quickly—as many as five generations in a season. The female lays her eggs in cotton buds, and the developing larvae destroy the seed and the cotton fibers.

❸ **Killer Bees, South America** Brazilian beekeepers imported bees from Africa to improve honey production in the 1950s. These bees tended to attack in swarms in response to even minor disturbances. Still, scientists hoped that breeding the African bees with tamer ones would produce less aggressive hybrid. It didn't work. The hybrid, known as the killer bee, started taking over honey bee colonies, moving steadily northward. They now live in the southwestern U.S. An individual sting is not deadly, but repeated stinging from a swarm can be. Researchers are working on ways to prevent killer bees from breeding.

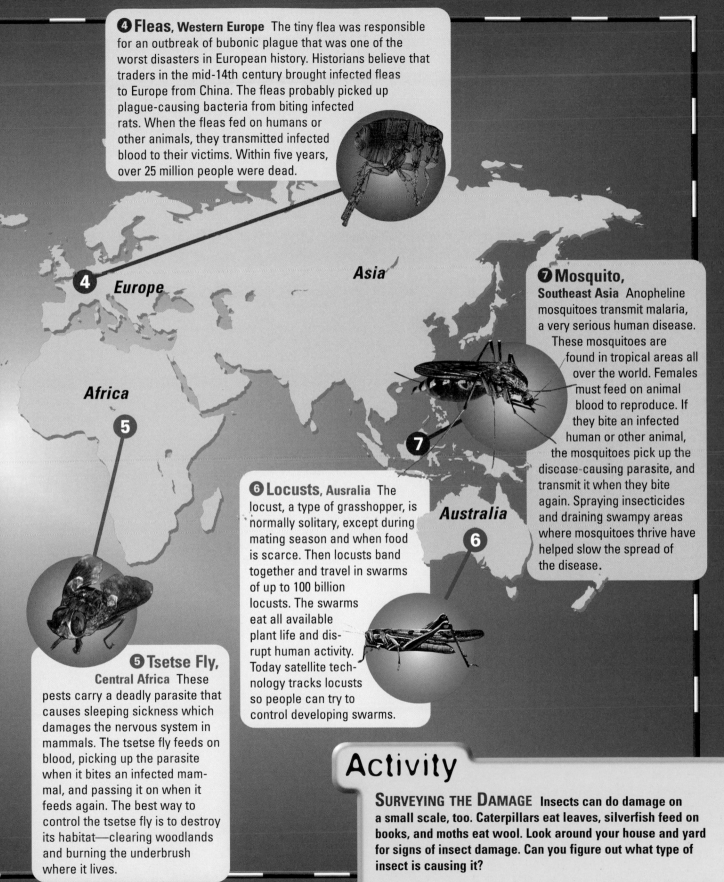

❹ Fleas, Western Europe The tiny flea was responsible for an outbreak of bubonic plague that was one of the worst disasters in European history. Historians believe that traders in the mid-14th century brought infected fleas to Europe from China. The fleas probably picked up plague-causing bacteria from biting infected rats. When the fleas fed on humans or other animals, they transmitted infected blood to their victims. Within five years, over 25 million people were dead.

4 Europe

Asia

Africa

5

❼ Mosquito, Southeast Asia Anopheline mosquitoes transmit malaria, a very serious human disease. These mosquitoes are found in tropical areas all over the world. Females must feed on animal blood to reproduce. If they bite an infected human or other animal, the mosquitoes pick up the disease-causing parasite, and transmit it when they bite again. Spraying insecticides and draining swampy areas where mosquitoes thrive have helped slow the spread of the disease.

7

❻ Locusts, Ausralia The locust, a type of grasshopper, is normally solitary, except during mating season and when food is scarce. Then locusts band together and travel in swarms of up to 100 billion locusts. The swarms eat all available plant life and disrupt human activity. Today satellite technology tracks locusts so people can try to control developing swarms.

Australia

6

❺ Tsetse Fly, Central Africa These pests carry a deadly parasite that causes sleeping sickness which damages the nervous system in mammals. The tsetse fly feeds on blood, picking up the parasite when it bites an infected mammal, and passing it on when it feeds again. The best way to control the tsetse fly is to destroy its habitat—clearing woodlands and burning the underbrush where it lives.

Activity

SURVEYING THE DAMAGE Insects can do damage on a small scale, too. Caterpillars eat leaves, silverfish feed on books, and moths eat wool. Look around your house and yard for signs of insect damage. Can you figure out what type of insect is causing it?

FORWARD MARCH

Army ants aren't like other ants because they don't build permanent nests. They swarm from place to place in search of food, eating until there's nothing left. Reptiles, birds, and even pigs and goats that don't get out of the way can be reduced to skeletons in hours.

NIGHTMARISH VISION

What would you do if a horde of army ants was headed your way? That's what the main character in science fiction writer Carl Stephenson's classic story, "Leiningen Versus the Ants," has to decide. Written in 1938, the story is fictional, but it paints a vivid, if exaggerated, picture of how army ants behave.

"Ten miles long, two miles wide—ants, nothing but ants! And every single one of them a fiend from hell." That's how a government official describes approaching army ants as he tries to persuade Leiningen to flee his Brazilian plantation. But Leiningen scoffs, confident his farm is safe because he has prepared two lines of defense—a ditch he can flood with water from a nearby river and a ditch he can fill with gasoline.

As the ants reach the edge of the farm, it appears that the waterway will to stop them, but then they start to march across on the backs of other ants. Leiningen's men spray them with gasoline and pelt them with dirt, but the ants keep moving.

Army ants carry ant larvae as they move from place to place.

One man struck with his spade at an enemy clump . . . [I]n a trice the wooden shaft swarmed with upward scurrying insects. With a curse, he dropped the spade into the ditch; too late, they were already on his body. . . . Wherever they encountered bare flesh they bit deeply; a few, bigger than the rest, carried in their hindquarters a sting which injected a burning and paralyzing venom.

TEMPORARY RESPITE

Leiningen's next move is to increase the water flow in the ditch, which sweeps some ants away and sends the others into retreat. Thinking the fight is over, Leiningen heads home. The next morning, he observes surviving ants gathering leaves, which he assumes are for food, but learns their real purpose when he sees the ants crossing the water on leaf rafts. As he rounds up workers to try to stop them, he witnesses a horrible sight.

Down the slope of the distant hill, there came towards him . . . an animal-like blackened statue. . . . When the creature reached the far bank of the ditch and collapsed opposite Leiningen, he recognized it as a stag, covered with ants. As usual, they had attacked its eyes first. Blinded, it had reeled . . . straight into the ranks of its persecutors, and now the beast swayed to and fro in its death agony.

Brazilian
army ant

LAST-DITCH EFFORTS

Turning his attention back to the rafts, Leiningen
raises and lowers the water level in the ditch,
drowning many ants. Then he makes a terrible discovery: While the men have been repelling ants on
one front, others have crossed the ditch elsewhere
and are swarming toward the plantation. Leiningen
must rely on the gasoline ditch. When the ants
start crossing it too, Leiningen sets the gasoline on
fire. Once the flames die down, the surviving ants
start across again. After this scene is repeated several times, Leiningen decides to flood the plantation. He makes a run for a dam on the far side of
the farm. The ants attack him, but he manages to
release enough water for the flood. His farm is
devastated, men have died, and Leiningen
himself is seriously injured, but finally the
ants have been stopped.

> [The water] swelled over ant-stippled
> shrubs and bushes, until it washed against
> the foot of the knoll whereon the besieged
> had taken refuge. For a while an alluvial of
> ants tried and tried again to attain dry
> land, only to be repulsed by streams of
> petrol back into the merciless flood.

ANT-ECDOTES

Some details about army ants show that Leiningen's
story is based on fact:

▶ Almost blind, army ants move by sense of smell.
  They release chemicals as they look for food.
  Other ants follow the scent and lay down more
  chemicals, which causes more ants to follow. The
  result is a massive army.

▶ Generally army ants eat other insects, up to
  100,000 a day. But if another creature gets in
  their path, the ants will usually eat it, too.

▶ Swarm raiders are ants that move in a wide
  group. They also have a painful sting.

▶ Some ants sacrifice themselves for the benefit of
  other ants in the colony.

▶ Marching ants will climb a tree, eating any
  animal they find on the way up.

Activity

THE WHOLE STORY As a class, read the whole story of
Leiningen and the ants. It can be found in the collection
Ghostly, Ghoulish, Gripping Tales, Franklin Watts, New York
(1983) or on the Internet at:
www.bnlhost.com/shorts/stories/lvta.html
Compare Leiningen's individual actions to the group
behavior of the ants. What would have happened if
Leiningen and his men had evacuated and waited for the
ants move on? Write down your arguments for or against
his decision to stay, then divide the class into two teams
and stage a debate on Leiningen's actions.

The Case of the FOSSIL FILCHER

CRICKET VALLEY EARTH HISTORY MUSEUM

FOSSILS

"Hello? … Hmm … Okay… We'll be right there."

Cass Hopper hung up. "That was Bea Honeywell, head of the Cricket Valley Earth History Museum. Claims an insect fossil collection was stolen. Let's go."

I'm Arthur O. Podd, Private Investigator. Cass is my partner. There's not a crime we haven't cracked.

12:30 p.m., Cricket Valley Earth History Museum

Cass and I climbed the steps to the museum entrance, scanning for clues. Cass spied a receipt on the top step.

"Someone had lunch at the Silver Fish Cafe. By the amount, I'm guessing that person ate alone," Cass observed.

"Could be a clue," I said. "We'll check it out later."

Ms. Honeywell met us at the door. She looked exhausted.

"I'm glad you're here!" she said. "We're about to open an insect exhibit—much of what we're showing is on loan from other museums, including the stolen collection!"

"Don't fret," I assured her. "Show us where you think it happened."

Ms. Honeywell took us to a room where two museum officials were unpacking shipments and recording information on a log. "This is Midge Magid and Shel Darkling, our curators," Ms. Honeywell said. "Mr. Podd and Ms. Hopper are detectives."

"When did these shipments arrive?" I asked.

"Yesterday," said Mr. Darkling, standing to his full height of six-feet-five, towering over Ms. Magid. "Feel free to look at the other exhibit items."

Mr. Darkling showed us some butterfly fossils. "These are from the Eocene epoch of the Cenozoic era—about 54 million years old. They are very much like modern insects."

"These fossils," added Ms. Magid showing us some earwig-like creatures, "are isoptera from the late Mesozoic era, about 136 million years ago."

"The missing fossils are of some of the earliest known insects," said Ms. Honeywell "They're from the Paleozoic era."

Cass got permission to review the shipment log.

"This must be it," Cass said, reading from the log. "Paleozoic insect fossils. 9:34 p.m."

"How late do you work?" I asked.

"Last night until about ten," said Ms. Magid. "We stayed late to prepare for the exhibit."

"Were you two the last to leave?" asked Cass.

"Yes," answered Mr. Darkling. "That's why I can't understand how the collection disappeared. By the time we arrived this morning, it was gone."

"Thanks," I said. "We'll be in touch."

As we got our coats, I noticed a mint on the floor. I pocketed it—it could be a clue.

"C'mon, Cass. Let's eat."

1:00 p.m., Silver Fish Cafe

We sat by the window. The food wasn't bad—fast service, too. All of a sudden, Cass spotted something.

"Look," she said, pointing across the street. "That store— Evolutionary—don't they sell bones and fossils from all over?"

"Yeah. Let's settle up and pay them a visit."

The waitress brought the check and a couple of mints.

"Mmm!" said Cass, popping one in her mouth. "Want one?"

"No thanks," I said. "I already have one."

1:45 p.m., Evolutionary

Evolutionary is a cool store. Cass and I were looking at a green beetle when the shopkeeper greeted us.

"We just got a shipment this morning. If you like bugs, take a look at these old ones," the man said, pointing to a collection under the counter. "These fossilized insects are older than dinosaurs."

"Look at the price," Cass gulped. It was five figures.

"Thanks," I said. "Interesting. But we've got to run."

2:15 p.m., Cricket Valley Earth History Museum

Ms. Honeywell met us in the lobby, looking relaxed.

"Good news!" she exclaimed. "The fossils were here all along. A curator just misplaced them."

Cass and I exchanged glances. "Are you sure, Ms. Honeywell?" I asked.

"Oh, yes! Mr. Darkling assured me that everything is back to normal."

"Just to be safe, we'd like to see the collection," said Cass.

"Of course," smiled Ms. Honeywell, leading us back to the same room.

We hung our coats with the other trench coats on the rack. Sure enough there were some fossils. Were they from the missing collection? I took out my magnifying glass to check them out. "Look, Cass." I gave her the magnifier.

"Hmm. These *look* like insects," said Cass. "They have hard outer shells, segmented bodies, multiple legs. Is this wormy thing a primitive caterpillar . . . or something else? And on this one—are those antennae or legs?"

Cass gave me another glance. I grabbed my coat, but as soon as I put it on, I knew it wasn't mine. It was too small and had a hole in the pocket. "Ms. Honeywell, I think you have a thief on your hands."

What was suspicious about the recovered fossils? Who does Arthur suspect is the thief… and why?

Use these clues to help crack the case.

❶ Photos of some of the recovered fossils

❷ Evolutionary store hours: 11:00 a.m. to 4:30 p.m.

Clues

Answer on page 32

prawn

crustacean

sea scorpion

shrimp

❸

The Silver Fish Cafe — Guest Check

| TABLE NO. | NO. PERSONS | SERVER NO. | CHECK NO. 2651 |

March 23, 12:01pm

1 cheeseburger $4.00

1 soda $1.00

total $5.00

Thank You - Call Again

GUEST RECEIPT

| NO. PERSONS | DATE | CHECK NO. 2651 | AMOUNT |

45740

MAKING

Extinction. What does the word mean to you? Wiped out. Forever. It can result from natural causes—a catastrophic event like an asteroid hitting the earth or something more gradual like climate change. Extinction can also be brought about by human activities such as unrestricted hunting, pesticide use, or industrial development.

March 10, 1994, San Pedro, California

The Palos Verdes blue butter-fly, also known as the PVB, lives only on the southern half of the Palos Verdes Peninsula in southern Los Angeles County. By the time the butterfly was dis-covered in the early 1970s, only a few colonies remained. Its food sources of vetch and deer weed plants were in dangerously short supply. Butterfly larvae eat only specific plants, so a threat to their food source threat-ens the whole population. For the PVB, a combination of urban development, weed-killer use, and possible competition from other butterflies depleted its supply of food plants.

After a sighting in the early 1980s, the PVB was not seen for many years and was believed to be extinct. But on March 10, 1994, a small colony was acciden-tally discovered in San Pedro, California, 25 miles south of downtown Los Angeles. The dis-covery didn't mean the PVB was alive and well—it meant only that there was a chance to save the species. To give the PVB a fight-ing chance, its habitat had to be restored. Enter Arthur Bonner. He's a former gang member who has devoted much of his life to giving the butterflies that chance.

Environmental Protector

Arthur Bonner grew up in south-central Los Angeles. He spent most of his teen years in and out of detention centers for commit-ting petty crimes. At the age of 18, he shot a man with a pellet gun and spent more than three years in jail. When he was released from prison at the age of 22, Bonner got a job with the Los Angeles Conservation Corps, pulling weeds to help to restore the habitat of the endangered El Segundo blue but-terfly. Bonner says, "After spending three-plus years in jail, I knew it was time for me to get out and live my life correctly for myself and my son, Aaron, who was born the same day I went to jail."

The El Segundo project ended, but Bonner's conservation career did not. Dr. Rudi Mattoni, a renowned expert on the butterflies of California and the Technical Director of the Palos Verdes Land Conservancy, hired Bonner as his assistant. They worked to promote the recovery of the PVB at the Defense Fuel Support Point, a naval fuel depot. The ultimate goal of the project is to reintroduce the butter-fly into all of its former habitats, but the task is not an easy one. To succeed, it will require the labor, cooperation, and support of many people and agencies.

Arthur Bonner works to restore a habitat in which butterflies can flourish.

A HOME

Rebuilding a Habitat

Bonner and Mattoni have had success in restoring the PVB's habitat and raising butterflies in captivity, but sustaining them in the wild is proving difficult. With almost 20,000 plants in the ground, the food supply is ample, but it hasn't yet attracted a large butterfly population. Mattoni speculates that "new threats" such as predators or parasites might be keeping the butterflies away. Identifying these threats and finding ways to combat them will, according to Mattoni, take several years. He stresses that it's important to "save the whole environment in which [the butterfly] lives."

Bonner is now the senior re-vegetation technician at the fuel depot lab. He works hard to make the best possible habitat for the blue butterfly and to restore the general plant community. And he is responsible for the day-to-day rearing of captive butterflies. Bonner starts his day at 5:00 a.m. He tends to the plants, pulls weeds, and makes sure the butterflies are safe from harm. It's hard work, but Bonner knows his dedication serves a good purpose. "Exploring the world of conservation has given me a great new outlook on life and the chance to show my talent and concern to everyone who shares the dream of saving what's left of nature in our communities," he explains.

After four years of work, Bonner and Mattoni won a conservation award from the National Wildlife Federation. Going forward, Bonner says, "My biggest challenge is to propagate as many butterflies as possible, from raising them from eggs to trying to introduce them into a good habitat on this facility." He and Mattoni have faced setbacks. "It's hard to figure out why things go wrong," says Bonner. "Is it the weather, or is something infesting the pupa while it's waiting to emerge? These are questions that always make a big challenge because there are always new things to find out about the diversity of life."

It Pays to Conserve

Butterflies are vital to the natural world. They pollinate flowering plants and serve as food to other insects and birds. When butterflies flourish in an area, scientists know that the ecosystem is healthy. If butterflies are dying, chances are the habitat is polluted. That means that many other insects and animals may be at risk, too. Through their efforts to save the PVB, Bonner and Mattoni are reminding us of the importance of butterflies in our world as a whole.

Palos Verdes blue butterfly

Activity

DANGER ZONES Dinosaurs and insects existed together on Earth, but dinosaurs are extinct and insects are thriving. Find out more about ecosystems back in dinosaur days (which ended about 65 million years ago) and put your data in a chart. In one column, list adaptations of dinosaurs and insects. In another column, give information about predators and prey. In a third column, list the factors that may have led to the extinction of the dinosaurs. Based on the information you have gathered, what conclusions can you draw about why insects were able to survive when dinosaurs could not? Did the dinosaurs' extinction have an impact on the survival of the insects?

BUG Bites

You probably call all insects "bugs," right? Who doesn't? But an insect is a "true bug" only if it belongs to a specific order of insects. Still, we use the term for all kinds of insects, including bedbugs and water bugs.

You Don't Say

What's behind these "bug" expressions?

Computer bug This phrase usually means computer trouble, but there is an insect that does "compute." It is an old tale that if you count the number of times a snowy tree cricket chirps in exactly 14 seconds, then add 40 to that number, you'll know the temperature in Fahrenheit. Is it true? Try it out.

Don't let the bedbugs bite Bedbugs *do* bite, but they don't just live in beds. They're not as common as they once were. Look for them behind baseboards or in cracks in furniture, as well as in beds.

Mad as a hornet If their nest is disturbed, hornets (a type of wasp) react by attacking the source of the disturbance and stinging unwelcome visitors repeatedly.

Snug as a bug in a rug Some insects like rugs but would rather eat them than cuddle up in them.

Wasp waist Women's fashions in the late 1800s featured artificially narrow waists known as wasp waists. In fact, wasps, bees, and ants are characterized by a narrowing where their abdomens and thoraxes are joined.

Busy as a bee Worker bees are indeed busy. They fly far from home to get food, visiting 50 to 100 flowers in one trip, and come back weighed down with nectar and pollen to feed the other bees.

Model Insects

Scientists look at insects to try to figure out how humans can do things better.

Natural coloring—Car manufacturers are trying to recreate the shimmering quality of butterfly wings for paint used on cars.

On the fly—Insect flight is the model for a small device that carries electronic sensors. Ideally the device will help the military gather information from enemy territory.

Artificial intelligence—To find efficient routes for telecommunications traffic in computer networks, researchers are looking at how ants and other social insects move. Scientists have also created artificial ants on computers, which are programmed to lay down pheromone-like trails as they travel from point to point on a map. By watching the ants' movement patterns, scientists believe that the ants will, after repeated trips, show the shortest route among the points on the map.

Food for Thought

Have you ever eaten an insect? Most of us have, whether we know it or not. Tiny insects get into flour supplies or into other types of food at processing plants. They end up in the food we eat. They're so small you can't see them. Do these types of insects hurt us? Not in small doses.

Early humans probably ate insects on purpose and many people still do. Some favor locusts, cicadas, and crickets while others feast on large beetles. In Mexico, a type of water bug is raised, just like cattle or chickens, for food.

People who eat insects are actually making a smart food choice. Many insects are higher in protein and lower in fat than beef. They're also easier to raise and take up less room than traditional livestock animals.

Sound Off

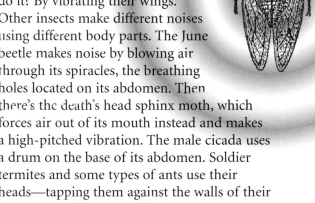

Flies and bees buzz. How do they do it? By vibrating their wings. Other insects make different noises using different body parts. The June beetle makes noise by blowing air through its spiracles, the breathing holes located on its abdomen. Then there's the death's head sphinx moth, which forces air out of its mouth instead and makes a high-pitched vibration. The male cicada uses a drum on the base of its abdomen. Soldier termites and some types of ants use their heads—tapping them against the walls of their nests, probably to warn of danger.

Wild Things

They're small, so we don't always notice them, but there are some really strange-looking insects out there.

❶ **Thornbugs**—As their name implies, these insects spend a lot of time on branches looking like thorns.

❷ **Ecuadorian Caterpillar**—Prickly? You bet. What's more, those spines are poisonous.

❸ **Male Giant Water Bug**—Those knobs on its back aren't for defense. They're eggs, which is unusual in the insect world—most species that tend their eggs leave the females to do the job.

❹ **Stink Bugs**—These bugs often have bright colors. Don't say you haven't been warned.

Bug Biographies

Recently, a college student noticed a praying mantis eating goldenrod pollen—an important observation, because before then, no one knew that that type of pollen was part of the mantis's diet. Much of what we know about insects we have found out from observation. Entomologists and others have been studying insects for centuries, but there's much, much more to learn.

Try your hand at observing and see what you can find out about the insects that are around you everyday. As a class, take a walk around the school yard and locate five different kinds of insects. Use a field guide to identify the insects and to read about their typical habitats and behaviors. Keep the behavior and habitat information handy for future reference. Divide into five research teams, with each team responsible for observing one of the five insects. Make a notebook for each team to use for recording observations.

Over the course of several weeks, about three times weekly, go out to the school yard and look for your insect. Try to make the trips at different times of day and in different types of weather. Make a record of

each trip in the notebooks. In your records, note the following data:

1. Date, temperature, time of day, and weather.
2. Where you found the insects. (If you don't find the insect, write down why you think you didn't see it.)
3. What you observed the insects doing.
4. Whether you found evidence (such as chewed leaves) of the insect having been in a particular spot.
5. Whether you saw more or fewer insects than the last time you made your observations.
6. Whether you saw any of the insect's predators.

Compare your observations to the information in the field guide. Is there anything different about what you've observed or is the insect behaving as described in the field guide? If there are differences, can you think of reasons why?

After several weeks, use the data in your notebooks to write up biographies of the insects. Include drawings of the insects in their habitats.

Ready for the ultimate challenge? Enter this or any other science project in the Discovery Young Scientist Challenge. Visit *discoveryschool.com/dysc* to find out how.

ANSWER Solve-It-Yourself Mystery, pages 26–27

Ms. Midge Magid is the thief. She stole the fossil collection from the museum and sold it to Evolutionary for an outrageous amount of money. Arthur and Cass deduced that Ms. Magid took the collection when she left the museum the previous night, and had time to sell it to the shopkeeper at Evolution when the store opened at 11:00 a.m.

Then she had time for a quick bite at the Silver Fish Cafe before noon. She lost the receipt and the mint through the hole in her trench coat pocket.

The replacement fossils are not actually insects, although they have some features in common (some are arachnids, some are small fish).

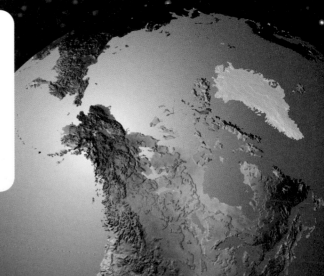